Maintaining portable and transportable electrical equipment

HSE BOOKS

©Crown copyright 1994

First published 1994

Applications for reproduction should be made to HMSO

HS(G) 107

HS(G) series

The purpose of this series is to provide guidance for those who have duties under the Health and Safety at Work etc Act 1974 (HSW Act) and other relevant legislation. It gives guidance on the practical application of legislation, but it should not be regarded as an authoritative interpretation of the law.

ISBN 0 7176 0715 1

CONTENTS

Introduction — **1**

Duties under the Law — **2**

Explanation of terms used — **3**

Controlling the risk — **4**

Use of the equipment — **5**

Construction of the equipment — **6**

Environment — **6**

Maintenance — **6**

Appendix 1 Legal requirements — **15**

Appendix 2 Summary — **17**

References — **19**

INTRODUCTION

1 This guidance note gives advice on the maintenance of portable and transportable electrical equipment, (see paragraph 12 for an explanation of these terms) which may be used connected to the fixed mains supply, or to a locally generated supply, and which could result in an electric shock or burn, or fire, due to damage, wear or misuse. The guidance covers electric drills, extension leads, portable hand lamps, portable grinders, pressure water cleaners, floor cleaners, electric kettles and similar equipment used in all environments, for example construction sites, factories, workshops, business and commercial premises etc.

2 It also indicates what the legal requirements for maintenance can mean in practice. It is intended to assist employers, employees and the self-employed, by describing what action can be taken to maintain the portable electrical equipment, over which they have control, in a safe condition, wherever it is used, and in so doing, to help them prevent danger arising. It is based on a straightforward, inexpensive system of visual inspection which can be undertaken by an employee of a business. Persons in control are, however, free to take other action or use alternative control measures that achieve an equivalent standard of safe maintenance.

3 The general principles outlined below are intended to cover all sectors of employment. More detailed guidance for particular sectors based on the advice in this guide may be prepared where experience suggests this is necessary. Separate guidance has been prepared relating to office, business and commercial premises and other low-risk environments, and also relating to the hotel sector (Refs 1 and 2).

4 Portable and transportable electrical equipment should only be used for the purpose for which it was intended and in the environment for which it was designed and constructed. It is unlikely that maintenance will remedy any situation where equipment is not being used for its intended purpose, eg table lamp used as a hand lamp, or in an environment for which it was not designed, eg a wet environment.

5 Major items of plant such as vehicles, cranes and generators are outside the scope of this guidance, as are electro-medical equipment and electrostatic spraying equipment, and equipment used below ground in mines for which there are special requirements.

6 Specialised equipment, eg information technology (IT) equipment (computers and printers), photocopiers, fax machines etc used in offices are not considered to present the same degree of risk, providing the lead and plug are protected from mechanical damage or stress. Movement, and therefore damage through being moved, is less likely to occur, and the equipment is often double insulated and used in a dry, clean environment with non-conducting floors.

7 Consultation is taking place to produce technical information for the use of those carrying out any necessary testing, and to assist small businesses and anyone who decides to carry out testing using their own staff. This gives advice and makes recommendations on what may be considered to be 'pass' or 'fail' conditions. In the past there has been an unnecessarily high failure rate for certain types of equipment often due to insufficient information or knowledge.

DUTIES UNDER THE LAW

8 There are legal duties on manufacturers and suppliers covering the initial integrity (safety) of new work equipment. There are general duties, covering the use and maintenance of work equipment, designed to ensure that it remains in a safe condition. Further details are contained in Appendix 1.

9 The particular legal requirements relating to the use and maintenance of electrical equipment are contained in the Electricity at Work Regulations 1989 (EAW). These apply to all work activities and place requirements on employers, self-employed and employees (subsequently referred to as duty holders), designed to control risks which can arise from the use of electricity. The Regulations require certain safety objectives to be achieved and do not prescribe the measures to be taken. This allows the duty holder to select precautions appropriate to the risk rather than having precautions imposed that may not be relevant to a particular work activity. For further information see the *Memorandum of guidance on the Electricity at Work Regulations*

1989 (EAW Memorandum) (Ref 3). Regulation 4(2) requires that *'as may be necessary to prevent danger, all systems shall be maintained so as to prevent so far as is reasonably practicable, such danger'*. This requirement covers all items of electrical equipment including fixed, portable and transportable equipment. The memorandum also gives further information on the meaning of reasonably practicable. Particular action that can be taken in order to maintain portable and transportable equipment, and thereby prevent danger, are described in paragraph 23 onwards.

EXPLANATION OF TERMS USED

Hazard

10 A simple definition of a hazard is anything that can cause harm if things go wrong (eg a fault on equipment).

Risk

11 A simple definition of risk is the chance (big or small) of harm actually being done when things go wrong (eg risk of electric shock from faulty equipment).

Portable and transportable

12 This is applied to equipment which is not part of a fixed installation but is, or is intended to be, connected to a fixed installation, or a generator, by means of a flexible cable and either a plug and socket or a spur box, or similar means. It includes equipment that is either hand held or hand operated while connected to the supply, or is intended to be moved while connected to the supply, or is likely to be moved while connected to the supply.

13 Though there are no universally accepted definitions of what is meant by portable or transportable electrical equipment, the definition given in paragraph 12 indicates the types of equipment covered by this guidance note, the supply to the equipment being at a voltage which can potentially result in a risk of fatal electric shock to any person, ie when it is more than 50V ac or 120V dc. Extension leads, plugs and sockets, and cord sets which supply portable equipment are classed as portable equipment since they operate in the same environment and are subject to the same use as the equipment they serve. Examples of portable equipment would be: tools and extension leads in the construction industry (high risk); grinders and hand-lamps in general manufacturing (medium risk); and floor cleaners and metal bodied kettles in offices (medium risk).

Note: The word portable is used subsequently to mean both portable and transportable.

CONTROLLING THE RISK

14 Nearly a quarter of all reportable electrical accidents involve portable equipment. The majority of these accidents result in electric shock; others result in fires, eg nearly 2000 fires in 1991 were caused by faulty leads to appliances. A major cause of such accidents is failure to maintain the equipment. The likelihood of accidents occurring and their severity will vary, depending on the type of electrical equipment, the way in which it is used, and the environment in which it is used.

15 A situation which can result in a high risk of serious consequences is the use of an electricity powered pressure water cleaner outside, powered by 240 volt electrical supply, with the cable trailing on the ground where it can be damaged by vehicles and other equipment, and where water is present. Damage to the cable or other parts is likely to result in the operator or others receiving an electric shock. Similar risks result when other electrical equipment such as drills and portable grinders are used in a harsh and sometimes wet environment such as at a construction site, where there is a high probability of mechanical damage resulting in danger. Lower risks result from floor cleaners or kettles which are generally used in a more benign environment, eg offices and hotels, but can be subject to intensive use and wear. This can eventually lead to faults which can also result in a shock, burns or a fire.

16 Such risks arising from the use of portable electrical equipment need to be assessed; factors to consider are referred to in paragraph 40. The risks can be managed and controlled by setting up an appropriate maintenance system including the measures referred to below. Because the consequences of an accident are so serious - potentially fatal electric shock, or fire affecting the whole premises - the maintenance system should be designed to be proactive, ie planned to prevent incidents arising, rather than reactive where action is taken following an incident/accident. The measures taken should be appropriate to the risk. Procedures will need to be carried out more frequently where the risk is high, eg on construction sites, and less frequently where the risk is lower, eg in offices. Much 'unauthorised' equipment is brought to work by employees, eg electric heaters, kettles, coffee percolators, electric fans. Use of such equipment should be controlled and it may need to be included in the maintenance regime (particularly the formal visual inspection) if its use is permitted. Use of equipment that fails a visual inspection should not be permitted.

17 The greatest overall reduction in risk will take place when the maintenance system is first put into practice. Thereafter it may take time to establish the appropriate frequency for any maintenance activity, as this has to be based on reviewing the measures over a period of time and readjusting them in the light of experience.

18 To assist those who are setting up a maintenance regime for the first time, suggestions are made as to intervals between both visual inspections and combined inspections and tests. The suggested intervals may be used as a guide by duty holders considering how to comply with legal requirements but every situation has to be considered in relation to the combination of the type of equipment, its use and its environment. Duty holders may therefore choose intervals that they consider to be appropriate.

USE OF THE EQUIPMENT

19 The reason for distinguishing between portable equipment and fixed equipment is that the electrical connections to portable equipment (eg the plug and flexible cable and its terminations) are likely to be subjected to, and more vulnerable to, physical damage and wear or harsh treatment, when in use, than is equipment which forms part of the fixed installation. The fixed installation is usually provided with a significant degree of protection against damage by the fabric of the building or fixed enclosure.

20 Equipment which is held by hand or handled when switched on will present a greater degree of risk because, if it does develop a dangerous fault, then the person holding it will almost certainly receive an electric shock.

CONSTRUCTION OF THE EQUIPMENT

21 For safety reasons some electrical equipment relies on the metallic (exposed conducting) parts of the equipment being effectively earthed (Class I type). If this earth connection is lost there is a possibility of the exterior of the equipment becoming live with a potentially fatal result. Anyone touching live metal will be in contact with electricity. Another design of electrical equipment (Class II) which includes double insulated equipment (marked ▣) is constructed with high integrity insulation and does not have or need an earth connection in order to maintain safety.

ENVIRONMENT

22 The risk of receiving an electric shock will be greater when the user of portable electrical equipment is standing on the ground outside or a concrete floor, scaffolding or similar which is a good conductor, than if standing on a wooden floor or dry carpet and not in contact with earthed metal work.

MAINTENANCE

23 Maintenance is a general term that in practice can include visual inspection, testing, repair and replacement. Maintenance will determine whether (a) equipment is fully serviceable or (b) remedial action is necessary. Routine inspection and appropriate testing, where necessary, are normally part of any overall strategy for ensuring that work equipment is maintained in a safe condition.

24 Cost-effective maintenance of portable electric equipment can be achieved by a combination of actions applied at three levels:

(a)   checks by the user;

(b) visual inspections by a person appointed to do this;

(c) combined inspection and tests by a competent person or by a contractor.

This should be followed up by management monitoring the effectiveness of the system, and action should be taken where faults are found, particularly where detected fault levels or types of faults are found repeatedly.

User checks (visual)

25 The person using the equipment can be encouraged to look critically at the electrical equipment they use and, after a minimum of basic training, visually check for signs that the equipment is not in sound condition, for example:

(a) there is damage (apart from light scuffing) to the cable sheath;

(b) the plug is damaged, for example the casing is cracking or the pins are bent;

(c) there are inadequate joints, including taped joints in the cable;

(d) the outer sheath of the cable is not effectively secured where it enters the plug or the equipment. Obvious evidence would be if the coloured insulation of the internal cable cores were showing;

(e) the equipment has been subjected to conditions for which it is not suitable, eg it is wet or excessively contaminated;

(f) there is damage to the external casing of the equipment or there are some loose parts or screws;

(g) there is evidence of overheating (burn marks or discoloration).

These checks also apply to extension leads and associated plugs and sockets. Checks should be undertaken by the user when the equipment is taken into use and during use. Any faults should be reported to management and the equipment taken out of use immediately. Management should take effective steps to ensure that the equipment is not used again until repaired by a person competent to carry out the task, (eg the defective equipment could be labelled as 'faulty' and its associated plug removed).

Formal visual inspections

26 The most important component of a maintenance regime is usually the formal visual inspection carried out routinely by a competent person. The majority of potentially dangerous faults can be picked up by such inspections, and the maintenance regime should always include this component. To control the risks and to monitor the user checks, a competent person should carry out regular inspections which include visual checks similar to those in paragraph 25 but undertaken in a more formal and systematic manner. Additional checks could include removal of the plug cover and a check made that a fuse is being used (eg it is a fuse not a piece of wire, a nail etc), the cord grip is effective, the cable terminations are secure and correct, including an earth where appropriate, and there is no sign of internal damage, overheating or ingress of liquid or foreign matter. The formal visual inspection should not include taking the equipment apart. This should be confined, where necessary, to the combined inspection and testing.

27 The competent person can normally be a member of staff who has sufficient information and knowledge, following appropriate training on what to look for and what is acceptable, and who has been given the task of carrying out the inspection. To avoid danger, competent persons should know when the limit of their knowledge and experience has been reached. Simple written guidance relating to this visual inspection can be produced, summarising what to look for, procedures to follow when faults are found and when unauthorised equipment is found in use. This can aid the persons carrying out the formal visual inspection and also users.

28 The inspections should be carried out at regular intervals. The period between inspections can vary considerably depending on the type of equipment, the conditions of use and on the environment. For example, equipment used on a construction site or in a heavy steel fabrication workshop, will need much more frequent inspection than equipment such as floor cleaners in an office. In all cases, however, the period between inspections should be reviewed in the light of experience. Faulty equipment should be taken out of service and not used again until properly repaired. If necessary, it should be tested.

The pattern of faults found can be used by management to indicate whether:

- the right equipment is being selected for the job;

- further protection may be necessary in a harsh environment;

- the equipment is being misused

in order to enable remedial action to be taken.

Combined inspection and tests

29 The checks and inspections outlined above will, if carried out properly, reveal most (but not all) potentially dangerous faults. However, some deterioration of the cable, its terminals and the equipment itself can be expected after significant use. Additionally, equipment may be misused or abused to the extent that it may give rise to danger. Testing, together with a thorough visual inspection can detect faults such as loss of earth integrity, eg broken earthwire within a flexible cable, or deterioration of insulation integrity or contamination of internal and external surfaces. Failure of insulation could result in the user receiving an electric shock with potentially fatal results. Periodic inspection and testing are the only reliable way of detecting such faults, and should be carried out to back up the inspection regime. Occasions when testing is likely to be justified are:

(a) whenever there is reason to suppose the equipment may be defective, (but this cannot be confirmed by visual inspection);

(b) after any repair, modification or similar work;

(c) at periods appropriate to the equipment, the manner and frequency of use and the environment.

30 The inspection carried out in conjunction with testing should usually include:

(a) checking of correct polarity;

(b) checking of correct fusing;

(c) checking of effective termination of cables and cores;

(d) checking of the suitability of the equipment for its environment.

31 Such combined inspection and testing should be carried out by someone with a wider degree of competence than that required for inspection alone, because the results of the tests may require interpretation and appropriate electrical knowledge will be needed. However this can often be carried out by a competent employee.

32 Persons carrying out **testing** of portable electrical equipment should be trained for the work they are to undertake. It is the employer's duty to ensure that they are competent for the work they are to carry out. Basically, there are two levels of competency.

(a) The first is where a person not skilled in electrical work routinely uses a simple 'pass/fail' type of portable appliance tester (PAT), where no interpretation of readings is necessary. The person would, of course, need to know how to use the PAT correctly. Providing the appropriate test procedures are rigorously followed and acceptance criteria are clearly defined, this routine can be straightforward.

(b) The second is where a person with certain electrical skills uses a more sophisticated instrument which gives actual readings which require interpretation. Such a person would need to be competent through technical knowledge or experience, related to the type of work.

33 It is likely that some combination of the actions in paragraphs 25 to 31 above will provide the most cost-effective way of ensuring, so far as is reasonably practicable, that portable and transportable equipment will be maintained in safe condition wherever it is used. The actions in paragraphs 25 and 26 are relatively simple. The more extensive inspection and testing described in paragraphs 29 to 31 can be carried out less frequently if the maintenance system includes formal visual inspections and monitoring of the user checks described in paragraphs 25 and 26.

34 Testing can be carried out at minimal cost where an employee has been trained to a suitable level of competence and provided with appropriate equipment.

Maintenance and test records

35 Although there is no requirement in the EAW Regulations to keep maintenance logs for portable and transportable electrical equipment, the EAW Memorandum does refer to the benefits of recording maintenance, including test results. A suitable log is useful as a management tool for monitoring and reviewing the effectiveness of the maintenance scheme and indeed to demonstrate that a scheme exists. It can also be used as an inventory of portable/transportable electrical equipment and a check on the use of unauthorised equipment (eg domestic kettles or electric heaters brought to work by employees).

36 The log can include faults found during inspection, which may be a useful indicator of places of use or types of equipment that are subject to a higher than average level of wear or damage. This will help monitor whether suitable equipment has been selected. Entries in a test log can also highlight any adverse trends in test readings which may affect the safety of the equipment, thus enabling remedial action to be taken. Care should be taken in interpreting trends where a subsequent test may be done with a different instrument to that used for an earlier test, since differences in the results may be due to difference in the instruments rather than indicating deterioration in the equipment being tested.

37 Records do not necessarily have to be on a paper system since test instruments are available which store the data electronically and which can be downloaded directly onto a computer database. Duty holders with large amounts of equipment will find it useful to label equipment to indicate that the equipment has been tested satisfactorily, ie has been passed as safe, and when the date for the next test is due. Otherwise individual items may be missed on consecutive occasions.

Frequency of inspection and of combined inspection and testing

38 Deciding on the frequency of maintenance is a matter of judgement by the duty holder, and should be based on an assessment of risk. This can be undertaken as part of the assessment of risks under the Management of Health and Safety at Work Regulations 1992 (Ref 5).

39 Paragraph 41 gives guidance that could be followed if a duty holder has difficulty in initially deciding how often to carry out inspection as well as combined inspection and testing, particularly where a maintenance regime has not previously existed. Alternatively, advice could be sought from a competent person who has the knowledge and experience to make the necessary judgement, eg manufacturers or suppliers of equipment, or relevant trade associations.

40 Factors to consider when making the assessment include the following:

(a) type of equipment and whether or not it is hand held

(b) manufacturer's recommendations

(c) initial integrity and soundness of equipment

(d) age of the equipment

(e) working environment in which the equipment is used (eg wet or dusty) or likelihood of mechanical damage

(f) frequency of use and the duty cycle of the equipment

(g) foreseeable abuse of the equipment

(h) effects of any modifications or repairs to the equipment

(i) analysis of previous records of maintenance, including both formal inspection and combined inspection and testing.

41 The table below is provided as a guide to the frequency of formal visual inspections and the frequency of combined inspections and electrical tests for portable and transportable electrical equipment in some typical types of business. The table gives suggested starting intervals when implementing a maintenance programme. Where one figure is given, this is a guide for anticipated average use conditions; more onerous conditions of use will demand more frequent inspections, and/or combined inspections and tests. Where a range is shown, the smaller interval is for more onerous conditions of use and the longer interval is for less onerous conditions of use. It is up to the duty holder, where necessary with appropriate advice, to assess the conditions affecting equipment, which may lead to potential damage and/or deterioration and which should determine the maintenance regime.

Suggested initial intervals

Type of business	Formal visual inspection	Combined inspection and electrical tests
Equipment hire	Before issue/after return	Before issue
Construction	Before initial use - 1 month	3 months
Industrial	Before initial use - 3 months	6-12 months
Premises used by the public, eg hotels	See other sector-specific guidance	
Commercial/office premises	See other sector-specific guidance	
General low-risk situations	See other sector-specific guidance	

42 In premises where portable electrical equipment is used by the public, and where a duty holder does not have direct control of the action of the public, inspection may need to be done much more frequently, but this can only be determined by knowledge of the likely risks and subsequently modified in the light of experience.

43 In many premises, eg health service, education, hotels and offices, more than one inspection and test regime may apply to different types of equipment for the following reasons. Some transportable electrical equipment in offices, which is normally supplied from a plug and socket, and is not handled or moved frequently, eg a table lamp, may not be likely to be subjected to mechanical damage. Thus, in a relatively benign environment, these conditions can be described as similar to those for fixed installations and the need for examination and test, ie the generally accepted interval is every five years for business and commercial premises. However, these conditions do not apply to all office equipment, and some frequently used items, eg floor cleaners, kettles, free standing electric heaters which may be likely to suffer abuse and damage, would need to be inspected and tested more frequently, until results can be studied and failure rates analysed.

44 After the first few inspections, the information obtained can be used to give an indication as to the intervals before further inspections are carried out. The same is true for combined inspection/testing. A low failure rate would indicate that the interval can be increased and a high failure rate that the interval should be shortened; see also paragraphs 35 to 37 on record keeping.

APPENDIX 1

Legal requirements

1 The initial integrity (safety) of new work equipment when first supplied is covered by:

(a) Section 6 (as amended by the Consumer Protection Act 1987) of the Health and Safety at Work etc Act 1974 which requires *'any person who designs, manufactures, imports or supplies any article for use at work or any article of fairground equipment:-*

   (i) *to ensure, so far as is reasonably practicable, that the article is so designed and constructed that it will be safe and without risks to health at all times when it is being set, used, cleaned or maintained by a person at work;*

   (ii) *to take such steps as are necessary to secure that persons supplied by that person with the article are provided with adequate information about the use for which the article is designed or has been tested and about any conditions necessary to ensure that it will be safe and without risks to health at all such times as are mentioned in paragraph (i) above and when it is being dismantled or disposed of ....'.*
   See HSE publication *A guide to the Health and Safety at Work etc Act 1974* (Ref 4).

(b) The Low Voltage Equipment (Safety) Regulations 1989 which require certain safety objectives to be met, including inter alia being so designed and constructed that protection against hazards arising from the electrical equipment, and protection against hazards which may be caused by external influences on the electrical equipment is assured.

(c) The Supply of Machinery (Safety) Regulations 1992 which contain a general requirement for protection against electrical hazards.

2 The general duties covering the use and maintenance of work equipment in addition to the Electricity at Work Regulations 1989 (Ref 3) are contained in:

(a) Section 2 of the Health and Safety at Work etc Act 1974 which requires *'the provision and maintenance of plant .... that are so far as is reasonably practicable safe ....'*.

(b) The Management of Health and Safety at Work Regulations 1992 (Ref 5) which require employers to make *'a suitable and sufficient assessment of the risks to health and safety of employees .... for the purposes of identifying the measures he needs to take to comply with the requirements .... imposed upon him ..... under other relevant law'*. Such a risk assessment should include risks arising from the use of electrical equipment.

(c) The Provision and Use of Work Equipment Regulations 1992 (Ref 6), require the employer (person in control) to select suitable work equipment (regulation 5) and to *'ensure that work equipment is maintained in an efficient state, in efficient working order and in good repair'*.

APPENDIX 2

Summary

Use this to check whether you are managing the risks from portable electrical equipment. You need to:

(a) have a system of maintenance for portable (and transportable) electrical equipment;

(b) have identified the portable electrical equipment that needs to be maintained and obtained information on where it is used and how. Decide what to do about 'unauthorised equipment' brought in by employees;

(c) have provided straightforward training and information for all users (including yourself) to help them carry out user checks;

(d) set up a formal visual inspection system;

(e) give the job to and train someone to carry this out;

(f) consider brief written guidance relating to the visual inspection, what to look for and procedures to follow when faults are found (and when unauthorised equipment is in use);

(g) decide on the appropriate frequency for formal visual inspection. (If records of visual inspections are kept, the findings can be reviewed and the records used to check whether these inspections can be carried out less frequently or need to be carried out more frequently);

(h) find someone to test equipment that:

  (i) is suspected of being defective (but this cannot be determined by visual examination), has been repaired or modified;

  (ii) is due for a combined inspection test (or has never had one at the start of a maintenance regime);

(i) ensure that the person has sufficient knowledge, training and experience as well as access to further information and advice where necessary;

(j) decide on appropriate frequency for testing where this is necessary;

(k) review records of test results and use to check whether tests need to be carried out less frequently or perhaps more frequently;

(l) monitor all the arrangements and ensure that follow-up action is carried out including a review of frequency of formal visual inspection.

REFERENCES

1. *Maintaining portable electrical equipment in hotels and tourist accommodation* IND(G)164L HSE 1994 ISBN 0 7176 0718 6

2. *Maintaining portable electrical equipment in offices and other low-risk environments* IND(G)160L HSE 1994 ISBN 0 7176 0719 4

3. *Memorandum of guidance on the Electricity at Work Regulations 1989* (EAW memorandum) HS(R)25 HSE ISBN 0 11 883963 2

4. *A guide to the Health and Safety at Work Act 1974* L1 HSE ISBN 0 7176 0441 1

5. *Management of health and safety at work* Approved Code of Practice L21 HSE 1992 ISBN 0 7176 0412 8

6. *The Provision and Use of Work Equipment Regulations 1992* SI 1992/2932 HMSO ISBN 0 11 025849 5

Printed in the UK for the Health and Safety Executive
C150   1/94